W9-DAH-182

Global Warming

What's That Got To Do With Me?

Global Warming

Antony Lishak

Smart Apple Media

This book has been published in cooperation
with Franklin Watts.

Series editor: Adrian Cole, Design: Thomas Keenes,
Art director: Jonathan Hair, Picture researcher: Diana Morris

Acknowledgements:
Bryan & Cherry Alexander/Alamy: 23, 31. Fred
Bavendam/Still Pictures: 8. Martin Bond/Still Pictures: 14.
Cape Wind courtesy of CapeCodTODAY.com: 12br. Bob
Daemmrich/Image Works/Topham: 3-4,15. Digital Vision:
13. Mark Edwards/Still Pictures: 21. Warren Faidley/OSF: 7.
Raimund Franken/Still Pictures: 9. Eric Gay/AP/Empics: 6l,
16, 28c, 32. A.Ishokon/ UNEP/Still Pictures: front cover b,
back cover b, 6c, 20. ITAR-TASS: 18, 28l. Mark Lynas/Still
Pictures: 24, 25, 30. Phil Noble/PA/Empics: 12l, 29.
Novosti/Topham: 19. Sipa Press/Rex Features: 1, 6r, 22,
26, 27, 28r. J. Tack/Still Pictures: 11.
Woodmansterne/Topham: front cover t, back cover t.

Published in the United States by Smart Apple Media, an
imprint of Black Rabbit Books, P.O. Box 3263, Mankato,
Minnesota 56002

Library of Congress Cataloging-in-Publication Data

Lishak, Antony.
Global warming / by Antony Lishak.
p. cm. — (What's that got to do with me?)
Includes index.
ISBN-13: 978-1-59920-037-8
1. Global warming. I. Title.

QC981.8.G56L58 2007
363.738'74—dc22   2006029890

9 8 7 6 5 4 3 2

Contents

So what?

Gases exist naturally in Earth's atmosphere. They help trap the heat of the sun that is so vital for all life on the planet. But over the last century, the level of these gases has increased, and Earth is becoming hotter. Scientists have named this process "global warming."

What's it all about?

The effects of global warming, including changes to weather patterns, are caused by gases in the atmosphere, such as carbon dioxide. These are called "greenhouse gases" because they act like the glass that heats air inside a greenhouse. The levels of greenhouse gases have risen as a result of the increased use of oil, coal, and gas. These fossil fuels release carbon dioxide into the atmosphere when they are burned.

Many scientists are concerned about the effects of global warming and argue that people must change their energy use to preserve our planet. But some scientists believe global warming is part of Earth's natural heating process, and not a result of human activity.

Personal accounts

All of the testimonies in this book are true. Some are first-hand accounts, while others are the result of bringing similar experiences together to create a single "voice." Every effort has been made to ensure they are authentic. Wherever possible, permission to use the information has been obtained.

Ask yourself

The testimonies won't tell you all there is to know about global warming; that wouldn't be possible. Instead, as you encounter the different views, think about your own opinions and experiences. This will help you begin to address the question: "Global warming —what's that got to do with me?"

Many scientists believe that global warming will lead to an increase in storms.

A hotel owner

Many believe that, as a result of global warming, ocean temperatures are rising. If this continues, it will have a devastating effect on sea life. Howard is an Australian hotel owner whose livelihood depends upon a unique marine habitat.

To the rest of the world the Great Barrier Reef in Australia is a beautiful natural wonder. But Townsville, where I live, and my hotel in particular, depends on the thousands of visitors it attracts each year. If the scientists are right, by 2050 the rising sea temperatures will have killed off the coral that lives on the reef. My kids should pack up and leave now.

A diver on the Great Barrier Reef.

Fact bank

■ The Great Barrier Reef, off the northeast coast of Australia, is the largest coral reef in the world.

■ High water temperatures kill off the algae that provide food for marine creatures and give reefs their vibrant colors. This process is called "bleaching."

■ In a survey during the summer of 2002, scientists found the effects of bleaching on the majority of the Great Barrier Reef.

This aerial view of the Great Barrier Reef shows only part of this huge marine habitat.

The warming waters are killing the coral. Even if global warming was stopped today, and that's not going to happen, it would still take decades for the deadly process to stop. As the habitat and food chains of over a thousand sea creatures disappear, some of the most beautiful life on Earth could become extinct. Forget about trying to find Nemo—he and all the other clownfish won't be there!

Ask yourself this . . .

■ Why should we care that a non-human environment is being damaged?

■ Can you think of other animal habitats that are being destroyed by global warming?

■ What could be done to stop the destruction of the reef?

A "green" cyclist

Global warming can seem like too much of a problem for one person to make a difference. Simon is someone who believes that if we all change our habits then there is hope.

Most people know there is a problem, but they don't believe they can do anything about it. They blame it on "governments" or "industry." But it's our actions that contribute to global warming, so we can easily do something to reduce its effects.

People need to stop wasting electricity because most of it is still produced using fuels that release greenhouse gases. One simple way to save energy is to stop leaving televisions, DVD players, and computers on "stand-by." People could also reduce carbon dioxide emissions if they reduced the number of unnecessary car journeys they make and tried walking or cycling instead. I cycle past hundreds of stationary fuel-guzzling vehicles on my way to work!

You could try taking a shower instead of a bath. Showers use a lot less water—so less gas or electricity is needed to heat up the water.

Simon believes people's habits must change.

Fact bank

■ Cycling is a pollution-free form of transportation and it's healthy, too!

■ Cars and trucks generate over 30% of the U.S.'s carbon dioxide emissions.

■ Televisions left on "stand-by" account for 23% of all TV-related energy use.

■ In the U.S., $3 billion is wasted each year by equipment on "stand-by."

Ask yourself this . . .

■ Carry out a survey to find out how many children in your class come to school by car. What proportion of them walk or cycle to school each morning?

■ How easy would it be to follow Simon's energy-saving suggestions?

■ Can you think of other ways to reduce energy use?

Reducing energy use helps to cut down emissions produced by power stations.

A rural activist

Wind is a possible carbon-free alternative source of energy that could help reduce carbon dioxide emissions. But Joyce doesn't think it's such a good idea. She believes that rural areas should not be spoiled by groups of wind turbines—called wind farms.

This sculpture is part of an anti-wind farm campaign.

What's the point of ruining rural areas for the sake of a few megawatts of so called "clean" power? If the future of green energy production involves more of these clumps of wind turbines, then I'd rather have global warming! These ugly eye-sores are the worst architecture in the world.

And, it's going to get worse. The global plan to reduce greenhouse gas

These protesters in the U.S. are campaigning against a new wind farm.

Fact bank

■ Renewable sources of energy, such as wind power, are considered to be the "clean" alternatives to fossil fuels.

■ "Clean" sources of energy do not produce any emissions or other waste products. They include tidal and solar power.

■ Some wind farms are built in the sea and are not visible from land.

■ A recent report concluded that 12% of the world's electricity could be supplied by the wind by 2020.

emissions by 5 percent by 2012 will mean building more of these awful wind turbines. Where are they going to be built? And, who do you think is going to pay for them? Taxpayers like me, that's who!

If you don't live near a wind farm then you just don't understand—they aren't as quiet as people would like to believe. Try lying in bed listening to 60 of them when the wind speed is fairly strong. There must be a better way to save the planet.

Ask yourself this . . .

■ How deeply do you think Joyce has thought about global warming?

■ What would you say if a wind farm was built near your home?

■ What would you say if a nuclear power station was built near your home?

■ Why do we need "clean" energy?

A car lover

Carbon dioxide emissions from cars are a major contributor to global warming. But for most drivers, it's much easier to say we should drive less than it is to do it. Gordon is a car enthusiast from Detroit, Michigan.

No one's going to tell me what car I can drive. I'm proud to live in the land of the free. When I get into my SUV [Sport-Utility Vehicle], I'm asserting my right as an American citizen to go anywhere I want at any time I please. And being so high lets me look down on other road users and, let me tell you, when they catch a sight of me in their rear view mirror they sure do make way.

The number one priority is my family's safety and that's what my vehicle guarantees. And what's the alternative?

Public transportation? Get serious! If you live in the suburbs, you just can't survive without a car.

I've heard people talk about the environment and how the fumes from my car are heating up the planet, but I just don't see the problem. Many scientists doubt there is such a thing as global warming. Who are we supposed to believe?

Car use is increasing throughout the world, not decreasing.

SUVs are owned by millions of people.

Fact bank

■ SUVs have large engines that are not fuel efficient.

■ A recent survey found that in one year, the average amount of energy used by an SUV is the same as leaving a light bulb burning for 30 years or leaving a television on for 28 years.

■ SUVs now account for over 60% of the vehicles owned by people in the U.S.

Ask yourself this . . .

■ This man's priority is his family's safety. So why does he drive a car with high carbon dioxide emissions?

■ Should SUVs be banned in order to force people to drive more fuel-efficient cars? What implications could a ban have for some people?

■ How could public transportation be improved? Think about the services in your area. What would encourage people to use them more?

A flood survivor

Many scientists believe rapid climate change has started to alter weather patterns across the globe. Letisha is someone who suffered from the devastating effects of a flood.

Hurricane Katrina turned my world upside down. No one realized how bad the flooding would be in New Orleans. My mom and dad took us to the Superdome when the water started to come into the house. We'd only packed what we could carry because we thought we'd be back soon. There were crowds of people sitting around with nowhere to go.

Thousands of families were affected by the flooding in New Orleans.

My dad is a local politician. He had been campaigning since the flood in 1993 for improvements to be made to the levees that control the river. He knew that the whole area was at risk because the city is built below water level. Hurricane Katrina caused waves to surge up the Mississippi River and break through a levee that was designed to keep flood water out of the city. But these levees weren't built for today's bad weather.

I think people had started to become too relaxed about the storms that hit the Gulf Coast. But where else could we go? It was our neighborhood—but now everything has changed.

Fact bank

■ In the 20th century, average global temperatures rose by 1.1 °F (0.6 °C). They are expected to rise by up to 10.4 °F (5.8 °C) by 2100, increasing the risk of extreme weather conditions. New Orleans is in a hurricane area and has always been at risk from flooding. Many scientists believe that the effects of global warming will increase these risks.

■ Despite the construction of the Mississippi River control system in the 1870s, the river caused severe flooding in 1927, 1993, and 2005.

Ask yourself this . . .

■ If the Gulf Coast region is flooded so easily, why do people live there?

■ What do you think could have been done to reduce the risk of flooding in New Orleans?

■ Do you think it is right to link this flood with the effects of global warming? Was it just a natural disaster?

Flood water from the Mississippi River caused widespread devastation.

Timber logging

Trees absorb carbon dioxide and release oxygen in a natural process called photosynthesis. The health of our planet depends on the great forests to filter the air for us. But what happens if they are destroyed?

A logger

There is a big demand in my country for wood, and we have to get it from somewhere. What is the problem? Russia has a very big supply of timber. Almost 60 percent of my country is forest. Some of those trees have been there for 500 years. We could cut down trees every day for the next 10 years and it would be like snapping a tiny twig from one of these giant pines. It won't make a difference. And, in time, these trees will grow back again.

A Russian logger ready with his chainsaw.

Fact bank

■ Russia is home to the largest forested region on Earth, with more than 55% of the world's conifer trees. Russia's forests absorb about 10% of global carbon dioxide emissions every year.

■ Global warming dries out forests, causing fires to spread more rapidly. Forest fires give off large amounts of carbon dioxide.

■ Some forests are managed to ensure that for every tree that is chopped down a new one is planted.

■ Every three years an area of Russian forest the size of California disappears.

■ Some forests are being planted to act as "carbon sinks." They are intended to absorb the carbon dioxide emissions produced by some industries.

A local woman

When I was young, the rivers were clean. Now the oil from logging machines flows in the water. But it's not just the logging—even more of the forest is lost to the fires that logging activities cause. All they care about is money. Of course people need wood for furniture and paper, but it has to be controlled or permanent damage will be done; not just to my land but to the whole world.

Russian timber waiting to be transported downriver.

Ask yourself this . . .

■ How important is it to control logging across the world? What more could be done to prevent the destruction of forests?

■ Wood is a material that can easily be recycled. What does your family do with paper and wood products that it no longer needs?

■ Can you think of other materials that could be used instead of wood?

An African farmer's son

Many areas of the world, such as Africa, suffer from drought. Some scientists think that global warming is making the problem worse. Here is the voice of a child from North Africa whose family is struggling to cope.

This child is carrying water from a well.

My father is a farmer and he says that for many generations our ancestors have learned to live with drought. But in the last few years, things have become worse. Now our land is turning into a desert.

We grow the food we need to live, but if there is no rain then we can't grow anything. It has not rained properly in my area for three years.

Fact bank

■ The United Nations says that one-third of Earth's surface is at risk of turning into desert. In Spain, 31% of the land is in danger.

■ By 2025, over 60% of farmland in Africa could disappear, along with 30% of Asian and 15% of South American farmland. Scientists disagree about how much of this will be caused by global warming.

My father says we will have to move to the city. The government doesn't want us to do that. They say there is no room. But what else can we do?

My grandfather blames the farmers who grew more food than they needed. He says they took all the goodness out of the soil. Some farmers burned down trees to make new areas for crops but that turned to dry dust when the rains failed again. He says we should stay and that things will improve if we care for the land. But that won't make it rain again.

Ask yourself this . . .

■ Do you think this family should stay, or go and live in the city?

■ How do you think the effects of global warming are making drought conditions worse?

■ Is global warming the only reason that areas like this are turning into deserts? What other causes can you think of?

Many areas of Central Africa have become desert wastelands.

An Inuit hunter

The most extreme effects of climate change occur in the coldest parts of the world. Here is an Inuit who depends upon the ice remaining frozen.

Global warming is melting my world. Year by year the permafrost that our whole region sits on is thawing into the sea. Areas that were once great expanses of ice have turned into mudslides. We depend upon the ice. It has been the source of our survival for centuries. Now it is thinning and

The Arctic ice pack is melting rapidly.

huge cracks make it almost impossible to navigate the hunting grounds that have been used for generations. Experienced hunters have been stranded as ice has floated away from the mainland and drifted out to sea. As the ice breaks so does the food chain that we are all part of. The seal colonies have reduced and there are fewer lemmings and foxes, which means that polar bears are losing their main source of food. Now we humans are considering leaving.

An Inuit hunter on his snowmobile.

Fact bank

■ When permafrost melts, huge amounts of carbon dioxide that were trapped in the soil below are released into the atmosphere.

■ Antarctica has lost 5,000 square miles (13,000 sq km) of ice-sheet over the past 50 years.

■ Arctic sea ice lost 43% of its thickness between 1976 and 1996.

■ The melting of the Greenland ice sheet would raise sea levels by approximately 20 feet (6 m).

■ Some scientists have warned that polar bear numbers could drop by more than 30% by 2050.

Ask yourself this . . .

■ How much should you care that a few people who live in a remote, cold part of the world might have to move somewhere else?

■ Who, if anyone, should take responsibility for the melting of the ice caps and the rising of the seas?

■ How might rising sea levels affect the rest of the world?

An island inhabitant

Melting ice causes the oceans to rise. This is disastrous for people who live on land that is just above sea level, as this island inhabitant is discovering.

Fact bank

■ Tuvalu is situated in the Pacific Ocean about 2,000 miles (3,200 km) northeast of Australia.

■ Tuvalu's highest point is only 13 feet (4 m) above sea level.

■ New Zealand has agreed to resettle 75 people a year from Tuvalu.

The sea has already damaged land on Tuvalu.

My island, Tuvalu, is likely to be the first state in the world to be submerged by rising sea levels. Scientists say it'll be uninhabitable by 2025, and by the end of the century it will have disappeared altogether.

Already, stretches of beach are being swallowed up by the waves and the saltwater of the ocean is rotting the roots of trees. Many crops have been destroyed, and we are dependent on imported canned food. Last year, without warning, the sea bubbled up

Waves crash against coastal homes on Tuvalu.

through cracks in the ground, tearing up the roads and flooding our homes.

My whole world is dying around me. It makes me feel sad and angry at the same time; sad that eventually we will have to leave our home, and angry because this has been caused by the actions of others who only look after their own needs and don't see the bigger picture.

Ask yourself this . . .

■ If you were an inhabitant of Tuvalu what would you do?

■ What could be done to help people whose land is almost at sea level?

A nuclear supporter

Jean-Paul works in a nuclear power station in France. He believes that the answer to reducing greenhouse gases is quite straightforward. The solution to the world's increasing demand for energy lies in the production of nuclear power.

Some day we will have burned all of the world's fossil fuels. Of course, it won't happen tomorrow, but it is going to happen. Who knows what damage fossil fuel emissions will have done to the planet by then? And our need for electricity is going to keep growing.

It seems pretty simple to me. We have to embrace "clean" nuclear power.

I know a lot of people have reservations about its use, but

French politicians celebrate the future of electricity production by nuclear power.

Fact bank

■ In 1979, mechanical and human error in the nuclear power plant at Three Mile Island, Pennsylvania, caused dangerous radioactive gases to escape. The building of nuclear power plants in the U.S. stopped.

■ A recent report shows that nuclear power is the second-cheapest way to generate electricity, at $0.044 per kilowatt hour (kw/h), after gas at $0.042. Wind power costs more than $0.095 per kw/h.

A nuclear power station in France.

■ France is the world's second-largest producer of nuclear power. More than 75% of its electricity is produced in this way. It is estimated that this has cut France's carbon dioxide emissions by 50% over the last 30 years.

■ A by-product of nuclear power is radioactive waste, which must be reprocessed or stored for hundreds of years. The world's largest reprocessing plant is in La Hague, France.

technology is so much safer today. There are organizations, such as the United Nations International Atomic Energy Agency (IAEA), watching over safety, too.

By the year 2050, there will be approximately 10 billion people living on this planet. All of them are going to need power. Nuclear power is the only reliable, immediately available energy source that could meet their needs in full. And the waste is controllable. There are no carbon dioxide emissions from a nuclear power station.

Ask yourself this . . .

■ How accurate is it for this person to call nuclear power "clean"?

■ How can nuclear power be seen as "better" than power generated by burning fossil fuels?

■ Given that accidents have happened in the past, how confident are you that "safe" nuclear power stations can be built?

What does global warming have to do with me?

Levels of greenhouse gases in the atmosphere are at their highest level in 420,000 years. The temperature of our planet has risen 1.8 °F (1 °C) in the last 500 years—half of that has happened in the last century. Many scientists believe that global warming will affect everyone on Earth. The debate about what to do is less clear. Should we use "clean" energy sources or nuclear power? Should we change the way we lead our lives to reduce carbon dioxide emissions? Or, is it too late to do anything about it?

Global warming quiz

Here are some questions you can answer to help you think about what global warming has to do with you. Compare your score with that of your friend's, then go back through the book. Use all of this information to form your own opinion about global warming.

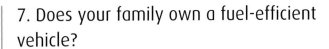

1. Do you always turn off lights when you leave an empty room?

2. Do you use energy-saving light bulbs in your home?

3. Do you turn off your television when it is not being used?

4. Do you take showers, not baths?

5. Does your family recycle?

6. Do you walk or ride a bicycle to school, only using the car when there is no alternative?

7. Does your family own a fuel-efficient vehicle?

8. Does your family buy products that are advertised as "energy efficient"?

9. Have you ever planted a tree?

10. Do you use public transportation when you can?

If you answered "yes" to less than 3 out of 10:

Read this book again and look at the Web sites on pages 30–31. Talk to your friends and family to discuss what global warming has to do with all of you.

If you scored between 4 and 8 out of 10:

You've clearly thought about what global warming has to do with you, but there is still more you can do.

If you scored more than 8 out of 10:

You know how your actions affect the environment. Talk to other people who didn't score as high about the other things they can do.

Web sites

The Web sites below feature more information, news articles, and stories that you can use to help form your own opinions. Use the information carefully and consider its source before drawing any conclusions.

www.greenpeace.org
Web site for Greenpeace International, with links to sections on climate change, nuclear energy, and forest protection, plus news, games, and campaign stories.

www.epa.gov/global warming/kids
Kids' Web site of the Environmental Protection Agency with climate information, history, games and movies.

www.worldwildlife.org/ climate
Climate change page of the World Wildlife Fund Web site with an introduction to global warming as well as conservation information and a section titled "What You Can Do."

www.iaea.org
Web site of the International Atomic Energy Agency, featuring news and stories from the nuclear industry.

www.reefed.edu.au
Web site sponsored by the Australian government featuring images of the Great Barrier Reef and facts about the plants and animals that live there.

www.unep.net
Web site of the United Nations Environment Network featuring information on changing climates around the world.

Glossary

Carbon dioxide – one of the gases found naturally in Earth's atmosphere. The buildup of carbon dioxide is believed to be causing a rise in air temperatures, which many scientists call global warming. Carbon dioxide is released when fossil fuels are burned to generate energy.

Drought – a long period without rain.

Emission – the action of something being released. When fossil fuels are burned, they emit carbon dioxide.

Enthusiast – someone who strongly supports something.

Food chain – a series of animals and plants in a "chain" that feed on each other and depend on those sources of food for survival. For example, seaweed ▶fish ▶seals ▶polar bears.

Fossil fuels – these are natural materials made from the remains (fossils) of creatures that are burned to generate energy. They include natural gas, coal, and oil.

Fuel efficient – using fuel with the least amount of waste or emissions.

Global warming – the increase in average temperatures throughout the world; thought to be caused by the greenhouse effect.

Green – a term used to describe someone who campaigns for the environment.

Greenhouse effect – the way Earth's atmosphere traps heat, like the glass of a greenhouse.

Greenhouse gas – a gas in Earth's atmosphere, such as carbon dioxide, that traps heat from the sun.

Permafrost – land that is usually permanently frozen, often to great depths.

Public transportation – the bus and rail networks that transport people from one place to another.

Rural – in the countryside, not in the city.

www.windpower.org/
en/core.htm

Web site of the Danish Wind Industry Association that includes news and information about offshore wind farms in Denmark.

www.nrdc.org/global
warming

Natural Resources Defense Council Web site with information about all aspects of global warming.

Every effort has been made to ensure that these Web sites contain no inappropriate or offensive material. However, because of the nature of the Internet, it is impossible to guarantee that the contents of these sites will not be altered. We strongly advise that Internet access be supervised by a responsible adult.

Index